CENTRALIA MINE DISASTER

UNNATURAL DISASTERS

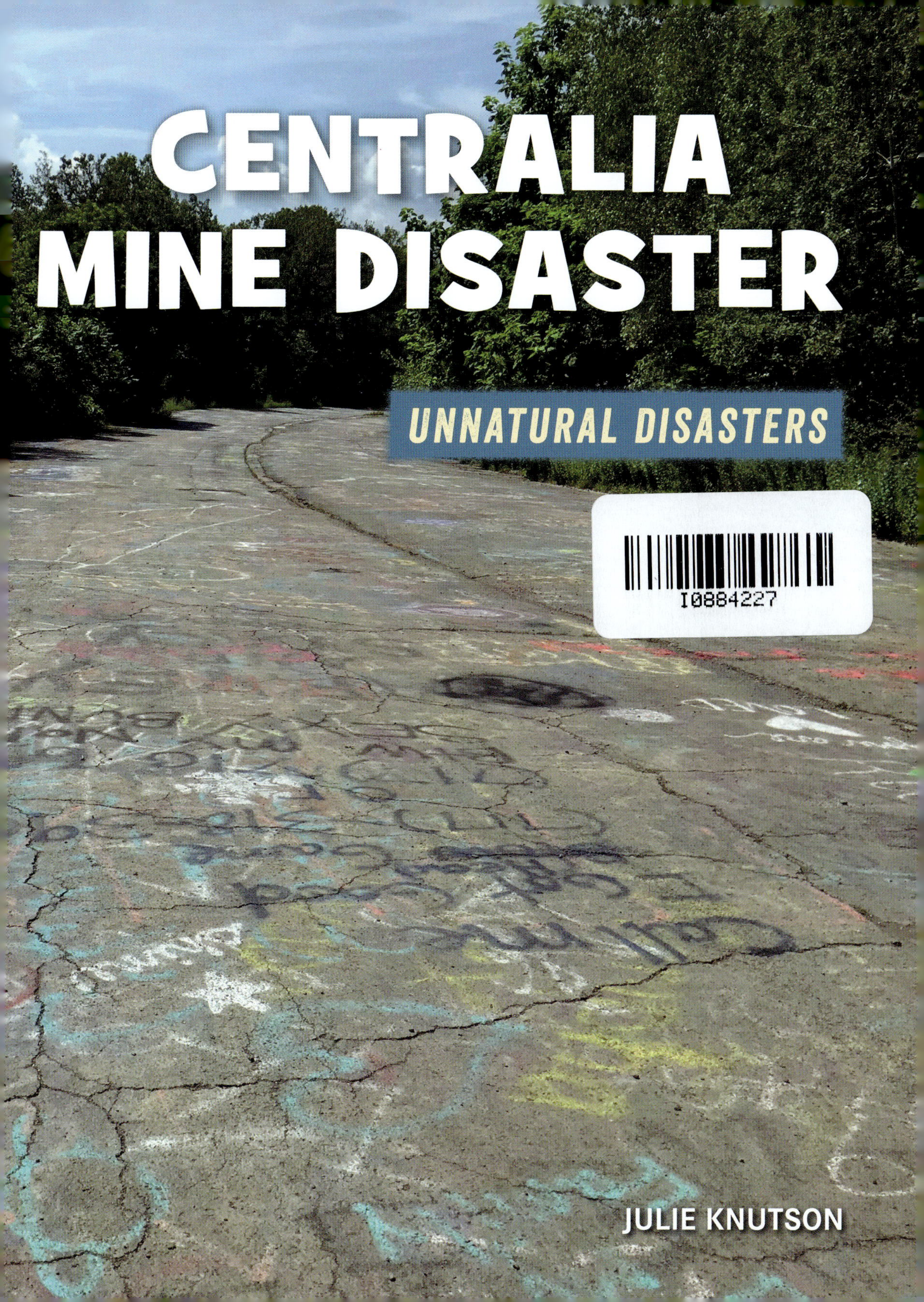

JULIE KNUTSON

Cherry Lake Press

Published in the United States of America by Cherry Lake Publishing Group
Ann Arbor, Michigan
www.cherrylakepublishing.com

Reading Adviser: Marla Conn, MS, Ed., Literacy specialist, Read-Ability, Inc.
Photo Credits: © 716.urbex/Shutterstock.com, cover, 1; © Everett Collection/Shutterstock.com, 5; © James Kirkikis/Shutterstock.com, 8; © RMC42/Shutterstock.com, 11; © Peter Gudella/Shutterstock.com, 12; © A. L. Spangler/Shutterstock.com, 14; © The Len/Shutterstock.com, 19; © urbanbuzz/Shutterstock.com, 22; © Alisa Metzler/Shutterstock.com, 26

Cherry Lake Press is an imprint of Cherry Lake Publishing Group.

Library of Congress Cataloging-in-Publication Data
Names: Knutson, Julie, author.
Title: Centralia mine disaster / Julie Knutson.
Description: Ann Arbor, Michigan : Cherry Lake Publishing, 2021. | Series: Unnatural disasters : human error, design flaws, and bad decisions | Includes index. | Audience: Grades 4-6 | Summary: "Human modification of the environment always carries a risk of accident and folly. Explore the causes and consequences of the Centralia, Pennsylvania mining disaster of 1947. Guided by compelling questions such as, "What led to this disaster?," "Who was impacted by it?," and "What changed in its aftermath?" the interdisciplinary content blends social studies and science. Ultimately, it pushes students to consider how humans can meet their need for resources in a safe, sustainable way. Books include table of contents, index, glossary, author biography, and timeline"—Provided by publisher.
Identifiers: LCCN 2020039961 (print) | LCCN 2020039962 (ebook) | ISBN 9781534180154 (hardcover) | ISBN 9781534181861 (paperback) | ISBN 9781534181168 (pdf) | ISBN 9781534182875 (ebook)
Subjects: LCSH: Mine fires—Pennsylvania—Centralia—Juvenile literature. | Mine fires—Social aspects—Pennsylvania—Centralia—Juvenile literature. | Mine gases—Pennsylvania—Centralia—Juvenile literature. | Coal mines and mining—Pennsylvania—Centralia—Juvenile literature.
Classification: LCC TN315 .K58 2021 (print) | LCC TN315 (ebook) | DDC 363.11/96223340977388—dc23
LC record available at https://lccn.loc.gov/2020039961
LC ebook record available at https://lccn.loc.gov/2020039962

Cherry Lake Publishing Group would like to acknowledge the work of the Partnership for 21st Century Learning, a Network of Battelle for Kids. Please visit http://www.battelleforkids.org/networks/p21 for more information.

Printed in the United States of America
Corporate Graphics

ABOUT THE AUTHOR

Julie Knutson is an Illinois-based author. In her spare moments, she enjoys investigating new places and ideas alongside her husband, son, and border collie.

TABLE OF CONTENTS

INTRODUCTION

With its huge population and coastal setting, Los Angeles, California, is a world away from the tiny, woodsy mountain burg of Centralia, Pennsylvania. But on March 15, 1981, this town of 1,200 demanded the attention of newspaper readers as far away as the other side of the country. On the front page of the *Los Angeles Times*, a headline screamed, "Flames Burn Underground: Centralia Fumes Over 19-Year-Old Coal Fire." A nearly two-decade-long mine fire—and the people who lived on top of it—couldn't be ignored any longer.

Why, after 19 years, was the world *finally* paying attention to Centralia?

It was because of a near-tragic event that had happened one month earlier. On Valentine's Day, 12-year-old Todd Domboski noticed a plume of smoke while playing in his grandmother's backyard. Curious, he walked toward it. Within seconds, Todd found himself sinking into the earth. He screamed in panic.

Coal mining has been an important part of the Pennsylvania economy since the mid-1700s.

A teenage cousin raced over to help. For 45 seconds, Todd struggled. Cold, wet ground closed in over him, first covering his knees, then reaching his shoulders and chin. But a tree root stopped him from sinking further, allowing his older cousin to grab his hand.

Miraculously, Todd was rescued from the 150-foot (46 meters) hole. Inside it, temperatures soared to 350 degrees Fahrenheit (177 degrees Celsius). Carbon monoxide readings were dangerously high. He was lucky to be alive.

Six thousand years. That's how long the world's oldest underground coal fire has been simmering in Australia. ***Fissures*** *in the ground near Burning Mountain (formerly known as Mount Wingen) still belch smoke and gas. In total, the fire has moved about 4 miles (6.4 kilometers) over the* ***millennia****. Today, it travels at a pace of 3.3 feet (1 m) per year.*

The event sparked national and even international attention. What started this fire? Why had it burned for so long? Would the people who lived there stay? Reporters from as far away as Australia and Sweden swept into the town with cameras and microphones, looking for answers to these questions. Centralia's story was featured on the popular TV news program *Nightline.* The town's residents were profiled in *TIME* and *People* magazines. The fire—and the conflicts it created in this small community—captivated the world.

Two sides faced off in the debate about Centralia's fate. There were those who vowed to never leave, and there were those who favored **relocation**. During a 2-year period, both sides presented their views. People wrote articles. They hosted parades, drafted **petitions**, and launched media campaigns.

Today, Centralia is mostly abandoned, with few structures dotting its overgrown roads.

Ultimately, in 1983, voters approved a **referendum** for relocation. Government funds were used to purchase homes, which were then **razed**. The fire, officials noted, would continue to burn underground until it exhausted its fuel. This could take possibly 250 more years.

This is the story of Centralia's unnatural disaster. It's the story of its human causes and of the people who suffered as a result. It's also the story of people who refused to be ignored, demanding answers and a safe, healthy place to live, grow, and thrive.

After the referendum, a few people refused to leave Centralia. Their houses—once part of a line of connected homes called row homes—were now freestanding structures in a mostly abandoned town. In 2002, the U.S. post office took away the town's zip code. In 2009, the 11 remaining residents were told they'd have to leave. Still, some refused.

CHAPTER 1

Before

Like oil, coal is a **fossil fuel**. It's the product of about 300 million years of Earth's work to mash, compress, heat, and transform ancient plants into hard rocks. Coal is buried in underground deposits called **seams**. Wedged between other types of rocks—like limestone or sandstone—coal's location is a record of its prehistoric origins.

Humans started to hunt for coal in the 19th century. As the **Industrial Revolution** seized the world, people looked for fuel to run steamships, trains, and machines. For decades, coal was king. It burned easily and provided energy. With 25 percent of the world's coal, the United States became a leading producer of this **natural resource**.

The Industrial Revolution brought about innovation and efficiency but also led to pollution and harsh labor conditions.

Anthracite, a type of hard coal valued for its high carbon content, is abundant near Centralia. But these deposits are locked deep below ground. In the 19th century, miners began to tunnel and dig vast underground networks to get at these layers of rock. Entire towns and industries sprang up around the pursuit. Centralia was one of them. In the second half of the 1800s, there were 14 mines operating in the area. By 1890, more than 2,800 people called the town home, many of them immigrants from Northern and Eastern Europe. Explosions, cave-ins, and falling rocks made the work extremely dangerous and dirty.

Mine tunnels like these provided oxygen to Centralia's underground fire.

While coal is still used as an energy source today, oil and natural gas began to replace it in the 20th century. After World War I, many new homes were built with electric, rather than coal, furnaces. Demand for coal fell. It mounted a comeback during World War II but plunged again in the late 1940s and 1950s. For towns like Centralia that were built on this resource, job losses, hardship, and shrinking populations followed. By 1960, the town was down to 1,435 residents.

The blaze that landed Centralia on the pages of newspapers and magazines in the 1980s started in 1962, with a fire at the town's dump. Across coal country, household waste was collected and heaped into old, clay-lined coal pits. Each year in May, Centralia's volunteer fire department would burn the trash to prevent odors, bacteria, and rats. Then, firefighters would quickly hose down the embers. But in 1962, the burn wasn't fully extinguished. And once the underground fire was burning, it was hard to put out. Abandoned mine tunnels provided oxygen, helping it to spread. It burned through everything in its path, including pillars and beams that provided structural support and prevented cave-ins.

How dangerous was coal mining in the late 1800s and early 1900s? Author Joan Quigley notes that one person died for every 35,164 tons of coal unearthed in Centralia's Schuylkill County in 1870. This fatality rate was three times higher than mining sites in England.

Lots of people had opinions on how to control the fire. Engineers toyed with different solutions, from digging it out to installing vent pipes to pull gases away from the town. The biggest attempt to stop

The disaster at Three Mile Island was one of the events that spurred a growing environmental movement.

the fire came in 1969. A trench was dug to intercept what was thought to be the fire's path. This would be filled with ash to halt its movement. But the fire was faster and dodged the attempt to cut it off. Soon, funds ran out. The fire still **smoldered** 65 feet (20 m) below Centralia.

By the late 1970s, the blaze crept closer to town. In the winter, patches of snow would melt on spots warmed from below. Fumes wafted through cracks in the road. In December 1979, a local service station was forced to close as 8,000 gallons (30,283 liters) of gasoline stored underground neared the boiling point. But still, people kept on with their routines. They adapted, installing carbon monoxide monitors in their homes to warn of immediate danger. The fire was something they simply lived with. It would take a major event to shift the balance of life in the town.

A Growing Environmental Movement

In the late 1970s, the environmental movement was branching off in new directions. Because of a series of man-made disasters, activists increasingly focused on how government and corporate decisions negatively impacted human health.

The 1978 Love Canal Crisis in Niagara Falls, New York, drew attention to the harmful effects of toxic waste on one community. Citizen-activists pushed not only for funds to relocate, but also for **federal** programs to clean up hazardous sites. The next year, a partial meltdown at the Three Mile Island Nuclear Power Plant near Harrisburg, Pennsylvania, led to widespread anti-nuclear protests. Both of these catastrophes were widely reported.

Centralians would have been well-aware of these ongoing battles waged by citizens to protect their right to health and well-being.

The Crisis: A Town at a Breaking Point

A few years before Todd Domboski got sucked into Centralia's soils, his father had these words for a reporter: "I guess some kid will have to get killed by the gas or by falling into one of these steamy holes before anyone will call it an emergency." That "kid" turned out to be his son. By coincidence, state politicians and local officials were touring the town on the morning of the accident. When emergency vehicles and news trucks descended on Todd's grandmother's house, Centralia's underground problem officially became an aboveground problem. It couldn't be ignored any longer. The threat was real.

After Todd's fall, residents debated what to do next. Two camps emerged. One group didn't want to—and couldn't imagine—ever leaving their community. They believed that the fire's path was moving away from the town. The other group wanted the fire dug out of the ground. If that wasn't possible, the government should **compensate** residents for their homes and assist with relocation. No one would buy a house sitting atop a mine fire. So without government assistance, most people would not be able to afford to move to a safer place.

Everyone in town knew that some kind of action was needed, and they were ready to push for it. Although the country had a new president whose administration placed much responsibility on states, residents sought help from local, state, and federal authorities. By mid-March, a group was going door-to-door to gather signatures to demand that Governor Richard Thornburgh visit the town.

Their petition got 1,600 signatures, some from people outside of Centralia. When Thornburgh visited in late March, he saw other activist efforts. Red ribbons were tied around the town's trees to represent the **red tape** that many Centralians felt was keeping them from getting aid.

Canaries in a Coal Mine

Ever heard the expression "Like a canary in a coal mine"? Typically, this **idiom** describes someone or something that warns of unsafe conditions. In the heyday of the coal industry, miners would bring caged canaries with them underground. Because canaries are more sensitive to dangerous fumes than humans are, they would die if exposed to toxic gases. If that happened, the miners knew it was time to evacuate the mine.

As underground fire raged, canaries became part of Centralia's households. A number of residents got them as pets to serve as living carbon monoxide monitors in their homes.

Using canaries to test air quality in mines dates back to the early 1900s, but the need for them inside homes was unique to Centralia.

At the end of April, about three dozen families and individuals formed the Concerned Citizens Action Group Against the Centralia Mine Fire. Some of the organizers had participated in the peace and anti-war movements. They were aware of the value of media attention, which could help pressure public officials and sway opinion. They needed to make themselves—and Centralia—visible.

Concerned Citizens planned a march in May. Children and adults carried signs that read "Centralia Held Hostage 6,940 Days by Mine Fire" and "Stop Talking, Start Digging!" and "Put *Out* the Mine Fire Before It Puts Us *Out*!" They made T-shirts bearing the words "Centralia: The Hottest Town in Penna." Many older residents of the community rejected these efforts and viewed Concerned Citizens as **agitators**. They preferred not to talk about the fire.

On May 19, 1981, Centralians voted on a **nonbinding** referendum. The question on the ballot was this: Should the town close if the fire can't be put out? In a news interview just days before, the mayor had predicted, "This referendum's going to be defeated.

In the fall of 1981, the Department of Interior bought and razed 27 homes in Centralia. These houses were in the fire's "impact zone" and had measurable levels of harmful gases. Owners were given money to relocate. But hundreds of others were left behind. Some wanted to stay. Others longed to leave but didn't have the finances to do so.

A lot of people have been here a long time. They're not going to pull up stakes now." But to his surprise, and that of many others, it passed. Sixty-eight percent of voters favored relocation. While residents' views had been measured, it was still unclear who would pay to move people out of the community. Wanting answers, Concerned Citizens took its fight to Harrisburg, Pennsylvania's capital, and Washington, D.C.

The should-we-stay or should-we-go debate simmered in the community for 2 years. In that time, Centralians were given carbon monoxide monitors for their homes. Every day, technicians ducked in and out of houses to record the air quality. People lost their privacy. They also lost their idea of home as a safe place. Those same technicians also monitored fire movements and underground temperatures through **boreholes**. By the spring of 1982, their readings suggested that the fire had moved closer to town and strengthened. By July, one temperature measurement reached 810°F (432°C).

Carbon monoxide is invisible and odorless. It is produced by burning fuel and, if not properly ventilated, can be very dangerous.

Tensions and division grew stronger. Concerned Citizens received **grant** funding to support its cause, but those who favored no change grew resentful. By mid-1982, this escalated to violence. Tom Larkin, the head of Concerned Citizens, got threatening phone calls. Another activist woke up at 4 a.m. to find his motorcycle shop firebombed. Several of the group's leaders resigned, citing health concerns and fears.

In August 1983, the community voted for a second time on a resolution to relocate. Again, the majority of homeowners voted in favor of moving. State reports of health concerns—ranging from bronchitis to cancer risk—confirmed what many people knew: their town wasn't safe. This report led to federal action. A bill set aside $42 million to buy the homes and relocate all of Centralia's residents. In 1984, demolition work began. Most people left, except for about 40 residents who refused—and they would hang on for the next two decades.

Comic book readers got a version of Centralia's story in the August 1984 issue of Action Comics. *Flames burned beneath the fictional town of Coaltown, but they were put out with the help of Superman. So in the alternate universe of comic books, Centralia managed to beat the fire.*

After

On a summer Sunday in 2020, cars wind their way over a freshly paved highway past what was once Centralia. There aren't visible plumes of smoke. There are no obvious signs of a fire burning underground. It looks like any other stretch of rural woodlands. But take a quick right or left turn, and you'll get a sense of the community that was once here. The footprint of a town—gridded streets and alleyways overgrown with grass and trees—remains.

In 1992, state officials seized control of the remaining homes in Centralia through **eminent domain**. But still, some people refused to leave. They sued the state in a lawsuit that lasted two decades. The legal battle came to a close in 2013. In the end, it was decided that the last residents of Centralia could live out their lives there.

While the town of Centralia is largely gone, it's far from forgotten. Books and documentaries continue to tell its story. Former residents still return for special events, like the unveiling of a time capsule buried in 1966. The Eastern Pennsylvania Coalition for Abandoned Mine Reclamation has been especially active in ecosystem restoration.

Today, there are 38 active coal fires in Pennsylvania alone. Hundreds of others are spread across the country and world. In China, it's estimated that 200 million metric tons of coal burn in underground fires each year. These uncontained blazes contribute about 1 percent of global carbon emissions and spew other toxins, like mercury, into the atmosphere.

The colorful graffiti on the road leading into Centralia was all that remained to remind people of the town's struggle. In 2020, that too was destroyed.

Volunteer events like cleanup days and tree plantings seek to beautify the landscape, improve animal habitats, and offset the region's carbon footprint.

Today, that move toward thinking about human impact can be seen in the landscape itself. Just beyond town, wind turbines tower above pine trees, churning out **renewable energy** as a clean alternative to coal.

A nearly 1-mile-long (1.6 km) stretch of abandoned road in Centralia became a tourist attraction. Graffiti began appearing on the asphalt and covered it by 2010, drawing curious travelers. To deter visitors hoping to leave their own mark on the spot, the state of Pennsylvania covered the highway in the spring of 2020 during the COVID-19 pandemic.

Research & Act

After Todd Domboski's fall, Centralia's Concerned Citizens used a tried-and-true method to demand government action: a petition. In March 1981, they sent this six-line message to Governor Thornburgh, along with a request that he visit the town:

> *"Dear Governor Thornburgh, Because of the danger to human life due to deadly mine gases emanating from the Centralia Mine Fire, we the undersigned urge you, the Governor of this Commonwealth, to visit this Borough of Centralia to determine first hand the gravity of the situation and learn that as stated in the March 16, 1981, Shamokin News Dispatch Editorial that … a declaration of a state of emergency is still in order."*

Research the history of petitions. When and where have they been used successfully in history? What features make a strong petition? What do petitions look like in today's digital world?

Next, think about an issue that you believe deserves more attention. It could be a concern at your school or in the broader world. Would a petition be a good way to raise awareness about it? What other tactics or strategies could you use to bring it into focus?

Timeline

- **May 1962:** The Centralia Mine Fire begins at the town's dump site.
- **1969:** Officials launch a major effort to stop the underground blaze, but the project ends when money runs out.
- **1979:** A local gas station owner measures the temperature of his underground fuel tanks at 172°F (78°C).
- **February 14, 1981:** Todd Domboski is trapped in a sinkhole for 45 seconds before being rescued.
- **May 19, 1981:** Voters in Centralia approve a nonbinding referendum to relocate the town if the fire cannot be contained.
- **August 11, 1983:** In a second vote, a majority of Centralians choose relocation.
- **1992:** Pennsylvania's governor declares eminent domain on the town; the remaining residents file a lawsuit to defy the order.
- **2002:** The U.S. Postal Service takes away Centralia's zip code.
- **October 2014:** Centralia's "time capsule," buried in 1966, is opened two years early after an attempted theft.
- **April 2020:** Pennsylvania state officials cover the "Graffiti Highway" to deter tourists during the COVID-19 pandemic.

Further Research

Bartoletti, Susan Campbell. *Growing Up in Coal Country.* Boston, MA: Houghton Mifflin, 1996.

Blakemore, Erin. "This Mine Fire Has Been Burning for Over 50 Years." *History.com,* updated April 26, 2019, www.history.com/news/mine-fire-burning-more-50-years-ghost-town.

Cheney, Jim. "Centralia: The Ghost Town Abandoned Because of the Fire Beneath It." *Atlas Obscura,* April 30, 2014, www.atlasobscura.com/articles/centralia-pennsylvania-2014.

Johnson, Deryl B. *Centralia.* Charleston, SC: Arcadia Publishing, 2004.

Mussari, Tony, and Stan Leven. *Centralia Fire, 1982.* PBS, 2016, youtu.be/ow4UERykRbg.

Glossary

agitators (AJ-ih-tayt-urz) people who shake things up

anthracite (AN-thruh-site) a hard type of coal with a high carbon content

boreholes (BOR-holez) deep, narrow holes dug in the ground for exploratory purposes, such as locating gas

compensate (KAHM-puhn-sate) to give money as a way of recognizing loss, suffering, or injury

eminent domain (EM-uh-nuhnt doh-MAYN) the right of the government to claim private property, with compensation paid to property owners

federal (FED-ur-uhl) the central, as opposed to local or state, government

fissures (fis-SHURHZ) narrow openings or cracks, usually long and deep

fossil fuel (FAH-suhl FYOO-uhl) a natural fuel such as coal or gas, formed out of the remains of living organisms

grant (GRANT) money given to a group for a specific purpose

idiom (ID-ee-uhm) a common expression or phrase

Industrial Revolution (in-DUHS-tree-uhl rev-uh-LOO-shuhn) a large social and economic shift toward machine and factory-produced goods

millennia (muh-LEN-ee-uh) periods of 1,000 years

natural resource (NACH-ur-uhl REE-sors) a resource found in nature, ranging from coal to wood to water

nonbinding (nahn-BINDE-ing) not official or requiring formal, legal action

petitions (puh-TISH-uhnz) letters signed by many people asking those in power to change a policy

razed (RAYZD) torn down

red tape (RED TAYP) business or government rules that create difficulties meeting goals or tasks

referendum (ref-uh-REN-duhm) a public vote on an issue or political question

relocation (ree-loh-KAY-shuhn) a move to a different place

renewable energy (rih-NOO-uh-buhl EN-ur-jee) energy sources like wind, solar, and hydroelectric power that do not run out

seams (SEEMZ) banded deposits of coal that can be seen in layers of rock

smoldered (SMOHL-durd) burned slowly, with smoke but no flames

INDEX